Sea-Level Nerve
(Book Two)

SEA-LEVEL NERVE
(BOOK TWO)

prose poems

James Grabill

La Grande, OR • 2015

Copyright © 2015, James Grabill

ISBN: 978-1-877655-90-6
Library of Congress Number: 2015950952

First Edition
December 1, 2015

Cover Art by Miriam Nelson
Cover Design by Kristin Summers, www.redbatdesign.com
Author Photo by Bill Siverly

Published by
Wordcraft of Oregon, LLC
PO Box 3235
La Grande, OR 97850
http://www.wordcraftoforegon.com
info@wordcraftoforegon.com

Member of Council of Literary Magazines & Presses (CLMP)

Text set in Garamond Premeir Pro
Printed in United States of America

Acknowledgements – Sea-Level Nerve - Book Two

The author gratefully acknowledges publications in which these prose poems originally appeared (sometimes in other form).

Aji Magazine (US): "In the Ambience"
Alchemy (US): "As Cherry Blossoms Open Above Us"
The Bitter Oleander (US): "Sayings as They Overflow," "Forsythia Yellow in Blue Grays," and "The Future of Rhino Horn"
Blunderbuss (US): "Improvisatory Jazz on the Sound System"
Caliban (US): "Dawn Roars Alive through the Cells" and "Will of the Mind Falling Asleep"
Calliope (US): "Pulse Point"
Carbon Culture Review (US): "Green and Going"
The Chariton Review (US): "Civilized Identity"
The Chiron Review (US): "Night Flight"
Cimarron Review (US): "Tuesday at Work"
Clackamas Literary Review (US): "City Rhino"
The Common Review (US): "California Burning"
The Convergence Review (US): "Empire of Cabbage and Herbs"
The Cordite Poetry Review (AUS): "The Intersection of Traffic and Light"
Dark Matter Journal (US): "In the Keeping of Light"
Eating the Pure Light (US, anthology): "Photograph of G.G.G." (as "Sunflowers")
Elohi Gadugi (US): "Forest Council," "Morning Stays Where It Goes" and "The Idea of Amplified Guitar"
Eunoia Review (Canada): "The Scent of Brown Rice"
Germination (Canada): "The Idea of Eating an Orange"
Green Mountains Review (US): "The Idea of 2039"
The Hamilton-Stone Review (US): "The Idea of Native Frogs"
Harvard Review (US): "Luminous Door"
Harvard Review Online (US): "Reasoning from Before" (as "Reasoning Song")
Innisfree Poetry Journal (US): "What the Wind Keeps Saying" (as "Belief")
The Istanbul Literary Review (Turkey): "Saucer Fungus on a Doug Fir Stump by a Black Beetle"
The Kentucky Review (US): "Being Suspended in Being" and "Branches Shaken by Light"
The Kerf (US): "Emerson at Night" and "Small Place"

The Laurel Review (US): "Classical Sousa" and "We Must Adapt"
The Lost Coast Review (US): "In the Approach to 2025," "Romney and the Nature of Light," and "The Scent of Afternoon Pavement"
Magma (UK): "Blind Spot"
Make It True: Poetry from Cascadia (US & CAN): "Iron Rails in the West" and "The Idea of 2020"
MiPOesis (US): "Questions for Applicants (Form B)"
Mirror Northwest (US): "November Yard Work" and "The Same Rain"
The New York Quarterly (US): "Waiting for Thunder"
The New Writer (UK): "The Unusual Medieval Occurrence" (as "Engendering Song")
Oxonian Review (UK): "Intelligent Machines"
Pemmican (US): "Bread and Apple" and "Lady Liberty in Red"
Poet Lore (US): "The Cat"
Poetry Quarterly (US): "Nonchalance"
Portland Review (US): "Hive of Road-Dusted Drivers"
ReDactions (US): "The Idea of Bees," "The Idea of Intrinsic Worth," and "The Rooster Is Nowhere If Not Awake"
Spittoon (US): "Unscrolled Historical Volts"
St. Petersburg Review (US): "Light Taking the Cells" and "Necessity of the Future"
Stoneboat (US): "Plumb Modern" and "The Idea of 2034"
Terrain: A Journal of the Built and Natural Environments (US): "A Few Miles into the Century," "Moths on the Front Screen," "Predicting Future History," "Pacific Trash Vortex," "Scarcities," and "The Idea of Heat 14° Higher Than Usual All July"
The Tribeca Poetry Review (US): "Civilized Theater"
Verse News (US): "Encircling Loom"
Visions—a Journal of the Arts (US): "In the Continuum" and "National Park"
Weber—The Contemporary West (US): "Grand Opening," "Inscrutable Northerly Drift," "Where All Morning the Climate Increasingly Turns," and "You Can Sit for Hours"
Willow Springs (US): "At the Ballpark of Exchanged Gnosis" and "Hieroglyphics in Wind from the Seeds"
Wilderness House Litereary Review (US): "Beauty in the Afternoon," and "Slowly Leaving Orbit"
Windfall (US): "Fir Trees in the Heavy Snow"
Written River (US): "The Ends of Desire"

Sea-Level Nerve: Book Two

Table of Contents

A Few Miles into the Century	13
As Cherry Blossoms Open Above Us	14
Beauty in the Afternoon	15
Being Suspended in Being	16
Blind Spot	17
Branches Shaken by Light	18
Bread and Apple	19
Breath in the Air Carries Weight	20
California Burning	21
City Rhino	22
Civilized Identity	23
Civilized Theater	24
Classical Sousa	25
Dawn Roars Alive through the Cells	26
Ecological Services	27
Emerson at Night	28
Empire of Cabbage and Herbs	29
Encircling Loom	30
Fir Trees after the Heavy Snow	31
Forest Council	32
Forsythia Yellow in Blue Grays	33
Grand Opening	34
Green and Going	35
Hieroglyphics in Wind from the Seeds	36
Hive of Road-Dusted Drivers	37
Improvisatory Jazz on the Sound System	38
In the Ambience	39
In the Approach to 2025	40
In the Keeping of Time	41
Inscrutable Northerly Drift	42

Intelligent Machines	43
Iron Rails in the West	44
Lady Liberty in Red	45
Light Taking the Cells	46
Luminous Door	47
Morning Stays Where It Goes	48
Moths at the Front Screen	49
National Park	50
Necessity of the Future	51
Night Flight	52
Nonchalance	53
November Yard Work	54
Photograph of G. G. G.	55
Plumb Modern	56
Predicting Future History	57
Pulse Point	60
Questions for Applicants (Form B)	61
Reasoning from Before	62
Romney and the Nature of Light	63
Saucer Fungus on a Doug Fir Stump	64
Sayings as They Overflow	65
Scarcities	66
Slowly Leaving Orbit	67
Small Place	68
Summer Day	69
The Ballpark of Exchanged Gnosis	70
The Cat	72
The Ends of Desire	73
The Future of Rhino Horn	74
The Idea of 2020	75
The Idea of 2034	76
The Idea of 2039	77
The Idea of Amplified Guitar	78
The Idea of Bees	79
The Idea of Eating an Orange	80
The Idea of Heat 14° Higher	81

The Idea of Intrinsic Worth 82
The Idea of John Rogers 83
The Idea of Native Frogs...84
The Intersection of Traffic and Light..85
The Pacific Trash Vortex 86
The Rooster Is Nowhere if Not Awake 87
The Same Rain..88
The Scent of Afternoon Pavement 89
The Scent of Brown Rice 90
The Unusual Medieval Occurrence..91
Tuesday at Work...92
Unscrolled Historical Volts...93
Waiting for Thunder..94
We Must Adapt 95
What the Wind Keeps Saying...96
Where All Morning the Climate Increasingly Turns 97
Will of the Mind Falling Asleep 99
You Can Sit for Hours 101
Author Bio.. 103

As I watch the bright stars shining, I think a thought of the clef
of the universes and of the future.

—Walt Whitman

We aren't separate from this planet; we are part of this planet.
Our minds may think that we are humans, that those are animals,
and those are plants, but I think many of us have come to think
that we're all part of one.

—Paul Stamets

As ocean shells, when taken
From Ocean's bed, will faithfully repeat
Her ancient music sweet --
Ev'n so these words. . . .

—Elizabeth Barrett Browning

The universe is shaped the way it would sound. . . . vibrations cause and are the reason for the Fibonacci ratios – shapes we find in parts of our bodies, in seashells, galaxies, crystals, sunflowers, orbits, eggs, buds, pines cones and in the unfolding of embryonic life. . . .

— Lorin Hollander

A Few Miles into the Century

At what point did sleep-swimming wake?

Where does being present begin and end?

Haven't North American nightcrawlers chosen to paw soft dirt by their holes to the other world, finding one another?

Has any act been other than collective?

Cycles at the root of heat, Bach crackling in integrity, shadows of Schoenberg thrown onto ground by sea-level—what lifts within cells may already know how a gull threads her flight in between stone in the sound.

What part of the whole would being exclude? Can the mind envision what presence portends?

Hammer those solar tiles down lightly, my boys. Tighten those photoelectric shingles under the sun. For wheeling within wheeling turns through current encircling, as pike at the bottom of the lake leave their imprints under giant galaxies in revolutions out of view of the eye.

Haven't we seen where carbon time-bombs in the air?

Oh Lord, woncha bring us a new Frigidaire?

For haven't we been where we haven't gone?

As Cherry Blossoms Open Above Us

When shuddering cherry leaves spoke, was it the wind? Or the old neighbors talking as if this were the country?

As if the days of untoward chemicals had not arrived?

Now cherry tree branches are thick with blossoms, each cell an individual life, each blossom one of the kisses on ancient walls of Hindu temples, a devotional Vedic hymn a woman sings like the day's forgiving.

The wood offers forgiveness, with so much flowing at once, so many wild working bees, that many people feel—how do you say?—re-vivified, restored to transnational value, perked up into solar aplomb, torched into current by the genome—?

Beauty in the Afternoon

The shot-putter steams ahead at a steady pace beside the road in the middle of the afternoon.

Running miles where Saturday traffic stops and starts under the canopy of sun and green overarching trees, wired for sound, she pushes on to the place that feels within reach.

A woman with long hair in the wind stands by the road, spinning a *Two for $5* pizza sign. Her slender moves reveal strength in her hands and arms.

She communicates pleasure that appears to come from another part of her life, or maybe from nearby green trees filled with sun, or from looking into drivers' eyes as they're looking back into hers.

In the sunlight and shade, she's magnetic. Her mouth and nose fit with her sync as she moves, older than some, still in her prime at the present, as being speaks with being.

At the fruit and vegetable stand, the person behind the counter stacks empty cases. When she glances up, her eyes deepen, focusing on someone who asks if the strawberries have sold out.

They have, she says, but you can still buy sweet onions and leeks.

The afternoon holds people close. In spite of disruption of the climate, abundance overflows. And it turns out existence is reward enough for effort made in our lives.

Being Suspended Within Being

We can't know longing in the world that led to our birth or damages done by the gods, only what the place has become.

Where night rolls out in its ten-ton trucks, consciousness is one tree in the forest we'd want to avoid clear-cutting.

At some point, single-celled beings reached into fin-making for their star-needled tidal swims. Isn't each second, hour, or millennium an exotic process of regeneration?

As honey-hive prairies carry on in back of gray-whale Pacific rains, a freezing hot whole range of slightest shifts falls in and out of sync.

If old-world heaviness frames the moment between opposing ends, isn't this a chance, if only a moment, to hear and be heard?

Where the mind is, however the place looks now, classical circulation feeds muscle memory and spontaneity. Disciplines convene in the smallest spectroscopic spike, as the rake of the spectrum returns to sea.

The warp of winds breaks up into fractal spin, gamma ray telescopes registering the beginning echo.

The way Earth looks from space, a place interconnected, smaller than ancestors thought, what happens next could suddenly appear.

Blind Spot

The rear-view mirror's blind spot burns with overexposures, heavy drapes on its windows darkening the past into coal. You turn your head to look, and a child leaps in front of your face, holding onto your hair while he tries to explain what you used to understand.

In no time, the backseat explodes with beautiful hounds that only want to reach the next trailhead where life begins and the dead world of waiting for teeth of the gears to slip to the bottom of their notches, forcing increased endangerment of little-known species, is erased.

Moaning and whining comes from the back, but you're driving through a Wyoming blizzard at rush hour. No, it's the bleeding glare of sunlight off gold plating of the professional sporting arena, where it floats between heaven and Earth, spitting out ocean ships through ten-story service doors.

You drive, and the blind spot is a city block glowering, smoking and blazing after future water shortages have forced climate disruption deniers onto the streets, where they've brought rifles, bazookas, and mobile launchers to the plaza.

But you can't see. You have to keep your eyes and the wheel well aimed, pursuing the business of post-carbon adaptation, half of your car burning, the other half taking a thermal.

Branches Shaken by Light

It's raining and cold in this part of the spectrum. The tree in the spine holds up, but what are we doing with our birds?

The past fills with emphasis, and empties each second before moving ahead. Arboreal rhizome makes core-spun seedlings sound. What happens next takes on necessity, as light reverberates within the spectrum.

A branch of the genome carries the shuck of mammoth breath.

Fallen leaves blow through layers of mushroom soils.

Overhead miles have their stretches of wing flying in further world.

Doesn't being here live in its forest? Doesn't the hour lengthen and shorten, depending on moss-green powers of ten?

A hundred-foot blue whale might draw up in a swell a few thousand lives from human thought. Breath fills lungs and overflows, guiding the Beethoven quartet in fluid time.

The warp of winds breaks into fractal spin where the moment's taken on weight, transporting parts of the future into the present.

A light of raw solar ore transplants energy in leaves and falls through Gauguin paintings into circulation of cells.

The future flies over on the shoulder of water. The compass eye aligns within cells, where a small cusp reflects the whole.

Bread and Apple

Her bread's the calm when light rain lifts from the drive.

Her apple will always be touched by lips, the way giant sunflower heads unwheel along transcendental fractals.

Her apple at its core remains an apple. It has already been an apple tree where a great horned owl brushes over history and the sky's a fabric steadily rewoven.

Her sourdough and oats may take rains working together on solitude as on elemental community, before yeasts are rising, daylight levitating, wild grains abandoning all but the next.

Where firs are daughters or sons of the talk within listening, wheels will be reaching her city and country for a taste stronger than the past, softer than mathematical proof.

Uncertainty glows in golden reeds by the lake. Parasympathetic ancestry can be saying, *we never seem to get enough good bread.*

Breath in the Air Carries Weight

Bela Bartok turns up alive, as waves in this nexus have gone buoyant with the inextinguishable chance to live here. The evening air reaches cells of the body, spreading through symbiosis, being within being, in the draw of Valeriy Sokolov's bow.

Strings resonate within Violin Concerto No. 2. Longing in the genome overflows. Night with its day feathers with imperative back through cells. In hands of Valeriy Sokolov, the Stradivarius resounds on its pilgrimage in translucence, quickening the sense that more is shareable of the whole.

The mammal fetus undergoes in-utero stages of evolution as air's being breathed, where transmissions reach in contiguous waves this nexus of what has become Bartok's Violin Concerto No. 2. As David Zilman conducts, Valeriy Sokolov stands before the orchestra, his bearing open, delivering virtuoso technique of a lineage of masters in a contemplative practice his alone.

Bela Bartok, through resounding moves, turns up alive in the concert hall as the concerto lives in this room. The solar plexus hears. The keyboard genome's played on by experience as shape's drawn into being through the torque of Valeriy Sokolov's bow.

Years ascend from old Europe rail yards, and descend within compass pours. The blanketing top of the collective sky sounds into space, as it shelters sleep and waking in the continuum.

California Burning

In forests, what a person owns she does not possess. What she owes is hers, and also not possessed. Even the iridescent flies that goad us into seeing them, mayflies lit on bark of trees, the ants maneuvering back through the forest brain—everything's vibration.

One day, a woman found herself driving a car through blazing California forests, though perhaps she was asleep. No, she was awake, but lost her thoughts. She lost her name as she lost great trees.

She lost Stellar's jays, rare warblers, great horned owls, ground squirrels, and so much more.

After she returned, she still had a name, but parts of her were gone, parts hard to define. When she looked for work, she knew she wanted a job helping people with what had been taken from them when they slept or were possibly awake.

Those she worked with gave her something she simultaneously found. She learned, through moving, ways to sink roots. She found the deeper her roots were, the more she could see, and the easier it was to move.

City Rhino

In the gray light, mid-morning, the rhino's still standing, taking in the rock and gray air.

As she yawns, the great grain elevator in back of her tongue drips, the air whistling in Kenyan with dragonflies over head-high grasses.

At the corner of the eye, ancient kinship branches. Gray calendars count with presence not numbers, or the Pacific rhino home stands naked, stopped in rolling alien scan.

Molecular light passes into blood-making marrow, but what's new? The close-by parallel world edges up, but who remembers?

Not much to be done. Water, food, inside, out, still arrive.

Boots sagging, concrete crumbling out of sight for decades, the rhino tips her horn in the vacant garage, the year the mechanic took off.

Where branches are kept impossible to see.

Afternoon sky cuts quick as it merges with the long story. The path returns but never left.

Civilized Identity

A person who lives without at least one animal and how the animal talks may be lonely for word from the cells.

As camera booms swing in, you can see dark wires fountaining, spitting archaic sparks from the future.

Private upkeep, the scent of long hair, a loaded dinner plate, the buoyancy of water in cells, you name it, will have been crackling awake under the sky.

Does wind follow what identity has been building? Are harbors invaded by so much ending or beginning? Can't a person be raised on river-street gruel by interdependent moms of evening rain and do all right?

Should the place be done up bright Mayan, with priest and hauler faces painted in a make-up of snakes and naked hope? After the great storm subsides, how much remains on the small tables of self?

A person who lives without at least one animal may be hungry enough to eat a complete horse. Or believing in human power, he may wield it.

Civilized Theater

Where shrieking *fire* without a fire in a crowded theater breaks the law, would being unable or ethically unwilling to shout once you've found the basement in post-industrial blazes also be a crime?

Should you pull the alarm before screaming? Is it polite to walk in front of the stage? Rather than *fire*, should you say *peace be with you*? If you know historical Latin, would this be the time to deploy it?

But would informing people simply be rude? Should you leave others in your row alone? Should you ignore those people who brought along cans of gasoline?

If through no premeditative intent of your own, you have sound reasons to think the foundation's melting, the basement roiling in flames, can you tell whether the process has been progressing slowly enough to warrant perfunctory legal action?

When part of the crowd, if you can no longer ignore the alarm, do you break the law? Could you still make a living after losing credibility? Could those who believe you?

And yet if you were to find your seat more comfortable the longer you sit, or you enjoy the film, as long as you aren't seeing curtains of flame where you are, should you ignore those who're barreling through as if you weren't there?

Can you sustain your civic politeness above all else, and doesn't your spiritual detachment feel liberating, as if no force could tilt your mortal coil, and the Earth were so tough nothing people might do could harm it?

Classical Sousa

The duende of Lorca's can be heard in Sousa—not so much from the schools, but the Marine band playing it two-thirds speed.

Ancestral long strides, rolling drums that wheel back cannon and the flag-covered fallen, arrive behind brass, whatever lifts or falls.

A college trumpeter who appears in red at half-time may be a sharpshooter who knows his aerial jazz and brass. He may be a wild songbird cutting loose, as if there's no other option.

But Marine trumpeters stand with both feet on heaviness of the ground—on solidness that doesn't move until it does, on the exact drop of the beat.

When the brass horns sound at once.

When parts fuel the whole, filling in one another.

The President strides onto the stage. The Marine brass resonates where breath's exchanged, where no sound's off.

Drums turn the wheel. Horns make Sousa talk and grieve. What has been, what has arrived and gone—hasn't been lost, when it has.

On the heart-pulse beat, what fills the horns will be streaming under speech.

Dawn Roars Alive through the Cells

The green wave crashes in slow motion through the summer.

The work week settles into sculptures of historical snow still falling from the towering downtown clock. At the root of strategic readiness remain whole legions of collaborating cells.

Sun radiates from the renaissance domes of electromagnetic waters and air in aesthetic unfolding.

Out of split-second auditoriums of whispering old-time capacity, in the roiling of stretched payloads aloud, through sleep solidarity between species that grew here from the first, the future sheds its mucosal croaks and snake skins, opening spiral amniotic eyes.

Nuclear projections of seed in bullfrog coughs out of the picture haunt encyclopedic balconies.

Rational classifications of steel ships maneuver between continents of future mothers.

Solar energy splits into color as intricate fractals middle the streets.

The unnamed world takes place in the intense 2050 spring.

Ecological Services

Appearances of the encyclopedia depend on contributions of native bees.

Roots of white oaks on banks of the river have a strong grip on what's underneath the visible world.

As research shows everything's more, may we remember it more.

The bear paw night holds down the salmon river.

Underground rain's drawn by the root through fiber to the veins and cells of leaves receiving the sun, slowly releasing water, purified, back into the air.

The beauty of flicker wing feathers is a display of collective cellular intelligence.

Not many animals alive in the wild are willing to let themselves go.

Vastness continues beyond orbiting, circling the origin, as attraction surrounds the genome.

The wide net the large porch spider continues to mend or reweave mid-air around the right spot expresses solidarity.

Emerson at Night

Neurological complexity involves cellular labors that contiguously retrofit the brain. Sophisticated faculties may be triggered at any part of the collective manufactory crawl.

At infinitesimal loads of matter, where each point is the center of the world, the question of intent may be mute.

Where one molecule approaches the urban boundary of another, do two subcontinents collide, both wheeling in a solar system, their force field haloes out of a Giotto Mother and Child emanating devotion, sobriety, and possibility?

Before reaching Boston, Emerson asks the driver to toss the reigns back onto the animal carrying them through the world.

The driver pulls the coach to the shoulder, and Emerson steps down through night air. He watches breath dissolve within overhead presence that protects the Earth with eternity.

The eye in the mind in the cosmos opens within and outside the picture we've framed for these bodies created by cells.

Everyone living owns forests and starlit air, whatever fiduciaries claim, where a person can ask, *Is this road I'm on the only one I could have been taking before I could wake, given the countless numbers who've lived for this time to come?*

Empire of Cabbage and Herbs

Spatulas slather paste over drying gashes on soldiers' arms. At supper, the ranks are rank in line at the soup pot. In the hospital tent by the rim of the white marble quarry, cabbage is spooned into mouths of infirm republic slaves.

Wild garlic grows under hot sun in Sicily. Its pungency drifts along stone afternoon roads through laboring heat where a citizen might use cloves to stay fit enough to serve. It is written aristocrats wouldn't touch it, only those from smaller houses out lop-hoeing summer fields, say.

Physicians wear crowns of laurel to protect themselves from thunderstorms and angers of anyone's gods. The Delphic Oracle chews bay leaves, ruminating in clouds of smoldering bay, laurel billowing her out into future vision.

The duke's cook adds handfuls of sage to fat-marbled venison, where sage was burned as velvet smudge. Vaporous fenugreek is ground into a poultice for unclogging an aunt's lungs, then to deepen desire when bathing. Cabbage and herbs follow their people into the future, where they've gone to live in the genome.

Encircling Loom

Amniotic drops of prehistoric dew on the canyon rim walls luminesce with light in the brain.

Complex otherness floats in womb-pulse swaths crossing the Pacific in adaptations of bodily cells.

Under blinding stars, electrical ancestral stories pour down through the small houses of losses and gain, where future scarcity looms and a dark-violet eyelash carries more weight than we know.

As the road goes out on its own, the root in a seed will decide. Where daylight drives the atmosphere, a shirtless boy swims in the sea of air. The cradle collides with shadow and magnetic lineage in current encircling turns.

As overflow sleep expands and contacts in the spectrum, the unfinished complex mind has an eye for complexity in the world.

Isn't this where separation from the whole grew opposable thumbs and set off on the road coming back?

What part of the whole would being exclude? What animals haven't loved and feared this air?

Fir Trees in the Heavy Snow

Snow had been showering this city, dream rain becoming moist and heavy snow, covering fir branches with gravity, as needles and their lift sag toward thinking the ground's where each flake's pointing.

The fir trees out back, thicker and less spontaneous, middle-aged, shield squirrels the way a person who works will help a son or someone outside a Burnside motel where most of the snow hardens for years, limbs holding miles of snow heavy as plaster.

We need to help one another, to keep the ocean currents flowing, where afternoon thickens with opaque weather, the snow of sunlight freezing, melting, sliding off in sweeps and sprays from heights as we stretch in position.

How long has heaviness covered us, sheltering what's been alive but locked inside the years following the war and accidents, when the sky was a roof that fell in on us? Forgive me for times I haven't understood you or known what to do, for the way I'll talk as if everything, even now, weren't okay.

Forest Council

Say in an ancient place, in an ancient time, we're on the council. Say we've assembled to apply very old language to the question of the future.

Our words are few and heavy, modified by moves of the arms and hands, looks of the eyes, and the back's position relative to the ground and fire, and the face, one person to the next.

Centers of words can be held by sense in the body. Beginnings of words can be punched, their ends rolled or delivered echoing what's close. You might hear whistles or clicks, pops or groans, snapping fingers, rattles of snake or gourd, thump of a log or walking stick on bone.

Words can be released instantaneously past smacked lips, topographic shifts of the face, in lift or fall of pitches when the old breaks through or presences conjoin, speaking through words.

The language is a mystery, when it talks into thought and the speaker hears its standing. We know the words well enough and progressions of phrase. We know this enough to respect it, the sacred trust of it.

So language works, as time spreads. Current time slows within. The ancient place we have no name for. It's where we live, and within us, not something else. It's so close what grows outside's in.

Forsythia Yellow in Blue Grays

Forsythias are easier to see in the flatlands. In the volcanic Northwest, unsprawled greens and brown-green browns shudder open, towering over stems of yellow flowers crawling with channels of sunlight.

Ground lifts into brown fir stands that crank open uncoiling ferns beneath them and lace the underground cooking its furnaces, fungi spanning more than a mile if left without basement diggings, without tunnels of architects who've taken written plans as sound.

The fir forests fog into breathing heights, the hill country morning witnessed as part of the mind, the fog boiling cool through yellow sounds in schoolhall lines, with sparks of the fog from machine hammerings that window the smallest industry.

Sheets of bird flyings all day are quilted between houses, blue jays alert mid-limb and talking a language lost if we hear, yellow branchings out from woven root-pulse embroidery on a blouse of fog, sewing gray sky bouldering ahead news and frequency at each point in gravitational bolting.

From the heat of roots, the juices of leaves, yellow on painted masks outlines a primitive brown and glows in the night museum hovering over day slightly sinking beneath day except in the brilliant yellow.

So billowing freighters houseled by misted old rains in the rivers, swimmers walking by old breakers cordoning away what has fallen out of reach, the canary's song more yellow than his feathers—the forsythia yellow flares through shops and the old bridge we use sometimes to deliver ourselves from what we believe should be here.

Grand Opening

'80s searchlights are still pouring up into the dark above town. Drowned out may be the ancient compass of stars, but who needs it? Sky-high beams swiveling, blazing, pass through the molecular remains of labor in the air.

The schooners have all reached land. Cities have already been peopled. Archconservatives are making the place free for the money. The store with no end is stocked with goods which are wonders of the world.

The ocean has kept going until reaching its destination. It has worked so hard to reach the shore, the apparatus set in motion may not be able to stop. But no fear. No one can help being drawn into the store with no end.

All the while, Civil Defense patriots with emergency-band radios they monitor from their kitchens keep mobile generators in a roar, pouring in gas, policing wires to the incandescent motorized washtubs the size of old cars. Steam in the summer evening swells around the great lenses at the smoking root of their beams.

The store with all anyone needs opens for the first time in the world. The nearby farmhouses are humbled. The governor of giant engines cuts the ribbon each time. Space-age doors slide open, and shut after you pass. Parking lots are packed with moving vans, as shuttle buses haul people in from long-term parking. Cargo jets keep landing in back, passenger jets from the East and West. Diesel trains blaring, roar in and disappear somewhere inside.

Junior and senior high schools merge everywhere you turn. Carmel chocolate for $10. Free coupons, 2 for 1, 1 for 2, you name it. Look like a model. Stand before mirrors, a wealthy ballplayer. Pray for the wealth of your nation, everyone one of us. Never fear appearing to be an animal again.

Green and Going

Before we could think, we believed in breathing. We knew what being mothered was, after living within heart-pump oneness and pitches of the mother's voice.

In womb-warmth, sea-sloshed when she moved, oneness would have been. Her breathing and waterfall pulse would have been.

Now you can be someone uncountable forcings hurl into the present. You can be someone in cold wind behind the garage when we were little.

What is the question held in check by a grove of old trees? Is this the place in which pieces of the past and future slip mostly out of view?

You can be someone driving a dark road without much time to go. In the quick, where we thought we were going would probably not be one place alone.

We move closer and further away within our means through contiguous inherited plumb. Where meaning has had a hand in shaping fiber, feather-strong pollinations spread.

Is this the story in which morning expands and contracts not exactly speaking our language? With conscious mind in the capillary rise, it's likely we'll miss wilder animals the longer they're gone.

Hieroglyphics in Wind from the Seeds

Grasses bend over anthills and the wavering doubt from brick streets, with pre-thinking's wind threading a tiniest edge, dipping us down in certain slow rhythm as if all this were the life.

The apple's birds will be more places than we learn, as the reaches, with so much forgotten, make bodily chords in simple breathing. Who we are, how we love, what is gone—may linger, but soon will be gone.

Berries form, and storms complete the sky. Wild open space shows the ant and rose where to cross in blossoming air, as moon scooping out part of the ocean lets the city suffer and sleep and exalt with cells and cradles.

A huge carved door that holds back English stories can open into a cream-colored horse ridden by birth multiplied by sun, as conjoined absences conceive. Stones will pour from a shovel of desire as if to say *Forgive us when we're wrong.* Many seeds go uncharted, where some will learn how to strengthen sunlight into food.

A struck gong already dissolved used to be doctrine, used to be long boats, where no-doctrine's contracting and expanding, attracting the young, the sea coast crashing, the sky inhaling much of the ongoing world.

Hive of Road-Dusted Drivers

From the overpass, you can see immense warehouses harbored at the edge of survival, where truckers have been backing their rigs up to docks, tenderly forming rows by their hives.

Inside the hives, individuals who've returned move their arms and trunks, motioning with their hands, dancing out directions in wheelings and turns, at times flaring up or fluting their wings as they hum and throat-sing what they've found.

A dancer's face can be grinning with pain, then grimacing in guitar rapture, depicting the known edge of a cloverleaf. Or falling flat with an ear to the ground can be listening to the dead, before launching out further in a direction.

Late at night, the decibel level rises as yellow jackets take pieces of flame along work curves in the middle of air. Or they practice signing the beautiful shoulder down the road or dragon-winged road on the small of her back, where the edge of shape is feathered in volts and paving oils and so on, in the place that repairs the sense of direction and burning readiness to be paid and pay out wild.

When hive-mind memorization winds down, the drones and workers store food in hexagonal chambers seen at the heights of chanting from before European settlements, everything running on photosynthesis where their labors map drives and detours specifically enough for some individuals to follow, but not all.

Improvisatory Jazz on the Sound System, Weekend C-Span on TV, People Inclined to Be Listening, Some Kind of Time Rounding the Clock

The physician who writes exquisite pieces announces, *There's nothing more intimate than having one's hands inserted in another's belly.* A progression happens, taking on life, as architectural construction continues in the empty place where everyone interconnects.

It turns out the book *First Man* describes walking on the moon, months of "zero gravity indoctrination," and the man who was first losing his father when he was eight.

A talking stream of notes flames up and resonates in molecular design making fractal Mandelbrot math possible to see, with chords settling their differences in the fractal complexity of plants as they grow.

The brothers Crude Oil and Coal, as close as they are to being history, as similar as all of us are, must know looming in the future is an invisible country where minds have cleared and all people are cared for.

While intent in this world means everything, what's done has been done, whatever reasons revolve around a tiniest enabling a larger, or vice versa.

A state senator from somewhere declares, *None of these liberties mean much after you're dead,* though who knows what he said before, as if meaning in the present hadn't come from the long-range future, as if a hard rain hadn't been drinking from rivers where light goes once it's done.

In the Ambience

Crest-cries spread in the flux of cellular hope. The small ribs of a feather fill with afternoon sun, exuding haloes of subatomic variables.

A slim chance mints comprehensions in the camera-phoned city, as heaviness presses down. Mallards descend quick to arteries of back-water sloughs. They lift into air and land as if it weren't anything, muscling the self into breath, new energy under the wing.

In a gallery, people peer into the photograph of an ancestor with a grip on the overturned boat. Holding the moment steady, the root of the sun cannot be seen from the Hubble.

Where the void lobs lightning into the gene pool, soaked as it is with passion and fear, offshore rain sharpens its draw where the story's out of our hands.

Urgency sweeps in with light, refracting down to a tiniest. Reeds at the lake have dragonflies flying luminous flutes of their spines. Of the viable composite in sleep-dive swoon, the lips fill in matter as in mind, riding at depths that break around us on waking.

If it does anything at night, the overpopulation expands. The hour ticks faster in poor villages on wrists of a river, unmasking as unequal distribution, bearing unfinished identity.

The scent of soil goes back to the earliest mammal mother, and out into the future rock-bottom shuck of mythical unthinking.

In the Approach to 2025

The next generation of people's names has been posted on the effects of civilization and the wilderness. The effort continues indefinitely. The old-faced future turns up the volume on ambience and miraculously can hear emptiness of the atmosphere call for a free-for-all fracking.

Archconservatives within the House vote to extend an additional absence of meaning. The pulp of it all steams between tooth-roaring crimson mirrors, the way colonies thrive where greed's being learned.

The long-term finished or not symphony of limitlessness refuses to end, rewarding memory with others' pain. One note to the next goes by in wind, as the future appears in lights mid-air, showing what to do and not do in what locations.

Silent witnesses in time may go by other names, the same as breath, which is an ongoing biological acceptance of elementary gases easy to overlook. The sky of storms builds further holes in the ground where the inexplicable sets up mansions.

The last money, the final ore, the only ropes, or you name it would know how much the more locomotive hungers ache, or how difficult it is to suffer risks to the future of others. It's certainly tough to follow the documentary when so many devices are working at once and the new war is further off in remote territory undergoing revision as we sit here.

In the Keeping of Time

We can be whole, part of the whole, refracted in the air with finches flying over in visible quickness.

Pieces of the past and future slip in between the least indivisible pulse and feather-strong mint conditions.

Up-right contiguous plumb resists the weight of mosses and mushroom underground, where questions held in check by a grove of historical trees have the same sky as breathing.

Accepting risks to the future of others we wouldn't know appears in the story of upper-floor suites, where the place no longer belongs to endlessness or only one of the species.

The Calvinist meat hornets should calm down, once they've eaten and taken enough.

A few holes in the ground can house the inexplicable, after snapping up common habitat, which can be easy to forget when locomotive hungers ache.

Because the ground must bear time as part of itself, its health weighs heavily on being here. Fir needles comb power out of fractal light.

Heart-pump waterfalls pulse mostly out of common view.

The melt of a peach in the mouth feathers slowly back into the sea.

Inscrutable Northerly Drift

The night and day sky continues to wheel in slow motion of the genome shifting with Celsius north.

At intermixed latitudes, the Pacific sea lions dive in swims beyond littoral brines, suspended in heaves, salt in the mouth of waters, the burn of bees in the forgotten and recollected, the quantum only hour as unfinished as all species.

Voices out of the future come with rain on back of drought, the inscrutable green where a chance exists, with dusts that fall in lobbies where forests have left their imprint.

Blood breathes out into blood, which breathes out into blood through miles in common, presence of species in common, where the extravagant overpeopling of hectares exists in common, Pacific island torches at the perimeter, the future drifting further north on its hungers.

Unexpected beauty we haven't named will take its time, however long the documentary footage resembles genetic complexity, where a feather can lift into thunderhead multiples and survive in jurisdictions of the unseen.

Intelligent Machines

If international equipment were required to go to school, would it grow two eyes and a human nose? Would the teeth in its mouth practice daily stewardship and carry on with a busy barnyard of mostly domestic inner organs?

Would sympathetic gears encourage one another to personalize intuitive transcendent forces? Would punch presses get the idea they're already larger than life? Would sad corporate switches decide they can never again go home?

Would certain door knobs realize they're meth addicts, and go off to an exotic country, to discover unfamiliar appliances willing to painfully kneel before highly advanced generators, after standing in line praying half a day?

Would tractor throttles learn later to connect their symptoms to unregulated petrochemicals they'd apparently been testing on their own? Would wrenches apply the principle of sanctity and practice being aware?

Would the brain of the thermostat keep abreast of the current stats, hoping to manage its attachments well? Would electronics understand the long haul and concept of *intrinsic worth* as it applies to all, gauges and regulators, condensers and panels in working order or not?

Would there be cases of suddenly older electromagnetic coils slipping into distress more than often over the words *if only*, as in *if only I'd been able* to hold another tight, in sweet slow-dancing to a hot loving ballad from Mo Town, *if only I'd experienced* more consciously the organic cocktail of desire and longing, bewilderment and adrenaline, after that first sense of body with body and the way it grew into a slow caressing eighth grade grind through what would have been street clothes?

Iron Rails in the West

A little presence of the future has been going past in a locomotive puff of barroom smoke in which Bengali tigers are forgotten.

Sparks from the tracks fire off into branches, where iron wheels roll the traffic of sleep headed for the middle of this century.

Was that the smoke of a barroom, or a bedroom in which the groom and bride have been watching the casino wheel revolve around inequity and abandoned US trains?

It's a sure bet the Reno at the striking end of a matchstick will be flaring up eventually into immortality, the way molecules circulate around incomprehensible bonds broken for new bonds formed.

Collisions of coal locomotives massively erupt into boreal forests blazing and roaring against the back kitchen window, where a stranger is looking in through tiger eyes.

Once the high-speed train has pulled to a stop at the future metropolitan station, people will be climbing down from their coal cars, flooding the marble neoclassical lobby with their many voices from 2033 still echoing when they reach back into us to change us.

Lady Liberty in Red

Her ocean banks fog up and smolder, where the ground forms under each step when she takes it.

She advances, and blue-shirted work crews arrive in phosphorescent vans before houses of the collective. Red lights blinking over hills of origin, the widowed brides of god whisper around a girl speaking with a friend.

Spreading deserts, dredged sea floors, oxygen missing offshore, forest hectares with earlier, hotter fires, the radio of ice broadcasting melts—widen the evening eye.

Earrings jangling, undiscovered fractions feathering, visionary young rooms smoking, individual cycles interlocking, remarkable bees threading the compass needle, the sweet vinegars of ruins and rice-burst continuum bearing, hatching—widen the evening eye.

Split-second dog-down hungers amass over miles rumbling and clanking on rails, as cost-benefit snowmelts gain gravity, primary as glacial grief.

Where the sun may have been conspicuously unclear about any intentions, the ability to stop swallowing Earth can be hung out to dry, as if people on all sides hadn't been working furiously.

A number still want to marry her, or become her, to walk openly like her, built out of gravity and waves, subdimensions and matter.

Light Taking the Cells

Light goes on reaching the metropolitan middle of the streets, passing through day as if it were space, as if atoms in the philharmonic midst were Vajrayana emptiness.

Folds of wheat peppered into orbits of bodily cells could be taking a sharp curve, pods and sheaths still floating out seeds, the cells working at night to weld animal sense to urgency, laboring to avert extinction.

What wasn't here when we first learned to speak, stillness within the continuum that continues to invent the wheel, long rains of antiquity reaching new swoons of bacterium, prismatic rings bubbling within sub-dimensions, has many projectors running at once, orbiting the central eye, blending characters this moment the civilization's unsettling.

The aerobic lungs and kidneys share these eyes, as the King's English clarifies *foul extraction* and *common good*, where luminous has been low-hanging moss on lanterned treks to the shore.

Light can be tasted as sun climbs into the sky in kernels of corn, rounding off what's seen, most of it tied to what hopes in a molecule.

Luminous Door

Hummingbirds against the black galactic intermolecular sky hover where matter itself has been space mostly and can be modified by the presence of suns in local orchards released, finally, not only from Genesis but Revelation behind its own cast-iron relief depicting kings and queens of forests with animals of soul.

The theater door, of course, has been carved out of hardened wood heavier than the Vaults of Hector or Heights of Ultraviolet in winds threading through solidness of sheltering rock with cellular deciduous root-sinkings tied into the elegant, however unintentional or primal.

Resourceful as it may have been to park by doors more open to what is still arriving from the periphery of a 1977 afternoon or some other world of a lord, it's hard to tell from here, upside down, if the last spray-painted symbols on the luminous exit to the unconscious will hear, *A hummingbird seen suddenly can be approaching infinity.*

Morning Stays Where It Goes

Root-reaching symmetry spreads through the air into wide leaves and orange blossoms of squash. Cellular time extends, expanding into the next limbs.

Future adaptation chances the slow pour of animals and plants into and out of Earth, where the sky's seen, and waking that would rather not sleep may be rolling on seriously revolving axles.

Unmitigated propagation unfolds. Nakedness begins in the continuum, with intrinsic worth of the finches and grasses, the standing maples and widening catalytic sweeps.

The ocean rolls through morning, the warmth of being in a room learning love beyond love, love within love.

As the sun tenders cells alive, life of the road breaks and heals over blends of the miles. The furnace that fires up Saturday resounds in root-smoldering presence echoing within the spectrum.

Swimming within indivisibility may unlock it. Water moves in the presence of light as light in the presence of water.

Moths on the Front Screen

The moth flies after transformation in arms of the plants. Ice melts in terrific splits and the truck of tissue and bone smoldering in wing-beat swells.

Far from the ocean, early night moths bring quickness to more of the children, as unmasked as anything is.

Going about their business in the mineral summer have been moths tendering the collective.

Through weight and infinitesimal lightness, in the workhorse ancestry of cells, the spectrum's packed into rock of day into night, the low roar of seawater weight asking, *What do you think will last?*

Unclassified Amazon moths will be drying their wings where the next leaves go under. Couldn't we have stopped at the root of hungers, before willingness to sacrifice the future for a little more now?

The moths land on the front screen between species. A few of the less settled bank off infrared Cheyenne energies, as a few choke on chemical mists.

Venus in the winds says, *This body floats on origin.*

A muskellunge lingers in cold pools at the bottom of water out of sight, but close enough to feel. A door blows open in the atmosphere from the long-term fault of *All this is yours.*

National Park

At the city park, elongated southern European wedding dances back in the genome experience soft brushes with poplars steaming with a finch song or filtering the air that enters a runner's lungs.

Long-remembered quick-tongued passages carried by exhalation through spiraling lengths of French horn arrive from marble pavilions and half-gutted motels under clouds of externalized future atmospheric costs.

Black Gothic half-tender rings around a girl's eyes can be drawing out the halfback from Great Plains religion in spite of intent, as a word on its own can be easily lost where sound seems to be going. And has it gone for good?

A presence of wolf out of nearby hills says, *Wait and then move. Move and then wait where you are.*

Step lightly, and run quickly.

Know where you are, as you know where you are.

Mark you, go where you go.

Brilliant green feathers fly their wing into the future tropical north, riding the buoyancy of gravity where probability falls in with unfathomable rains throughout history drying out in what's to come, where the unseen world's said to be, as unmasked as anything may be.

Necessity of the Future

A rhinoceros brushes past the back of the neck of civilization.

In the downpour, raw naked as a birthday, dispossessed Lear could not employ his plate. Poor Tom—him's crying out loud with Lear hopping hot and cold as your Roman Legion.

Not many animals alive in the wild are willing to let themselves go.

May no history ignore surgically removed bird entrails my smart cat placed by the door.

Swallowing hope may be easier where mountain gorillas would paint no prey on cave walls back in bison France. The mystical brother of Tecumseh still stands, where dreaming goes.

Where art comes close and works for subtle sense, all may be more.

If work shows all may have been more, may we remember it more.

Unheard chords may exist outside the options of sound, as mint soils of a slim chance operate in the back-roar.

Impersonal bees still sharpen their direction on what isn't loneliness or sleep leached into the uncommon wild, where what melts may have unfrozen before our eyes forever.

Night Flight

Inches in front of me, a woman with soft hair leans her seat back and touches her overhead light off in the dark sky.

The night is a vault through which we're passing, preparing for the death of someone close, or less of more species, or for the end of this stretch of 19th century assumptions giving way to the new.

When people thought more in soliloquies, what kept them warm when the rain chilled bones of the chest as well as nerve?

It's clear this jet isn't an Army plane, delivering us to a desert on the other side of the planet, each of us hauling gear, each with a liver and spleen to protect, but ready to occupy whatever parts of the map we're told.

Now the plane's a wing that lifted from a Midwest hospital, air tubes hissing overhead, attendants wheeling in a crash cart, ruins in flames beneath us, billowing columns of smoke over marks on the ground where jobs were lost, and lives, and ground.

After some time we're tunneling through shafts of midnight coal as saxophone players are drilling down, their horns lifting them through city blocks, progressive licks building into and out of shifts of chord.

In our atoms, we're flying to Austin, civilians walking the aisles, the voice of the pilot around us suddenly, astonishingly calm, as if he has forgotten where we are.

We can see out a window, to blood in the fog blinking, insinuating the reach of the wing. We know the morning will reach us, that the sun will appear at the unclear horizon, as much of the life of the planet has its blazing yellow infinity in the bees there.

Nonchalance

On two-lane Taylor's Ferry, a crow walks over to the deli container tossed onto the center line.

Before a next car passes on the busy side, the crow pivots on her high heels, her black cape sweeping down her back into wings, and walks calmly in the right direction far enough away.

After the car, she steps back to give it a poke with her beak. Nothing scrambles or falls out of it. Another car approaches, so she turns, half-strutting as if she were an officer in uniform, certain she can take care of herself.

You can't tell me she's not at home on this road, in this part of the city, whatever she may have imagined when she first left the training program of her beautiful mother.

Centuries pass through human quarters, as even preachers feed a regular gruel of *us and them* to the greater unloved and undone. Crows, in the meantime, navigate the system and mostly remain above it.

November Yard Work

Barbells clang from the neighbor's garage. And the wind sounds as if someone is letting air out from inside the trees.

They look like it, tired, losing their yellowed maps.

No, those are old letters from overseas forces blowing across the lawns, some from those whose listening is now sumac, whose living wages is our life, and we are not just folks here walking around entirely, definitely not these legs and arms and nails on fingers only, or hair in the wind or beards scattered on the body.

Definitely not just this brain the skeleton carefully carries like a serving person lifting a meal up the stairs to the Indonesian throne.

Now afternoon glows from walls as the brain cools into ancient cave paintings.

No, into figures only calculus could cast into long-ranged functions.

But we're out, listening to the birds beneath our lives, working on further into the fall, asking for simple understanding, for Calvinist forgiveness, yes, of course, and for regard, the kind that heals from inside birth that brought us to this planet.

Photograph of G. G. G.

In the photograph, my grandfather holds the stalks of twelve-foot sunflowers and is smiling, like a man who has given himself to beauty.

At the beginning of speech, his great great granddad's Swiss-German has sunk into English, the way an orange carries Bach in its diamond seeds.

We wanted the orange to be sweet and the sun inside it to know how to heal a woman's breast when she needs it.

Ten-thousand farm years pass, as cottonwoods tower over a cricket.

A harpsichord plays backwards into ancestors.

Did we ever have enough love?

Plumb Modern

Once where we thought we were, we were already going on in small bands, long before science rocketed out through time and parachuted into sea depths of mothers and fathers back through millennia.

Now so many of us live, are alive, breathing in this living place, that we're overflowing, more anonymous, collaborating and squaring off on ground where we're nearly identical.

Isn't this the place where people have trouble keeping up with much of what the collective has learned, where you can still practice magical thinking and, through hard work, almost believe it?

A sparrow sings the sun up into the sky. Gravitational mountains lean on the orbiting core beneath us. The aurora borealis harp of solar winds draws a bow over genetic strings.

When complex otherness fans out through whale-eye winds, what remains on the small tables of self?

The ease of moving which unfinished species have depended on can be heating up in mineral wildness, whispering *greenhouse*, where finch wings, as we sit here, continue to evolve through the being of finches and long-range Apollonian pollination.

Predicting Future History

The future history of recent events may well transpose underlings and substitute intent for conditions in the face of known extractions. It turns out only one species has been writing it.

Noticed will be extraordinary individual wealth with accumulations of impoverishment related to the transactions and consolidations that seem determined, perhaps bound, to pursue business as usual underground in lodgings as meet class requirements of comfort and attendance of serving people.

The latter will be stone-serious staying in studio apartments, surviving on meager rations, where the great future history of recent past survival of identity believes it independently exists without need of dissection tables or names for fabled microbes.

If the anti-Jeffersonian-liberal-education crowd has its way, identity will win social recognition, whether or not undertakers will have abandoned the old practice of attempting to placate any of the closer individually bereaved or shocked.

Why spend the final valuable moments of time decorating the next corpse with its own resemblance, in the future world experiencing the effects of possession of the means and ends by those long ago sequestered in comforts of their own underground?

Finding beauty in a broken world is acknowledging that beauty leads us to our deepest and highest selves. It inspires us. We have an innate desire for grace. It's not that all our definitions of beauty are the same, but when you see a particular heron in the bend in the river, day after day, something in your soul stirs. We remember what it means to be human.
— Terry Tempest Williams

I think the word "humus" has such power, because I think humanity comes from it, humility comes from it, humidity comes from it—everything that gives life and creates our humanity comes from it. So, even though it might look a bit strange, but I think creating organic farms and organic gardens is the single biggest climate solution, but it's also the single biggest food security solution.
— Vandana Shiva

It's as if there are two kinds of resonance. Sometimes it's as if the muscles in my heart were all violin strings, taut and tuned, and when a book sounds a tone, the same vibration trembles in me. So there is that kind of resonance, the sympathetic vibrations of a reader and a writer in perfect tune. But then, there is the resonance of the bell, which rings only because of its emptiness.
--Kathleen Dean Moore

We are living in a time where the myriad devastating crises facing the planet call for nothing less than a planetary transformation in consciousness— one that many believe is already underway. The burning question is how we might awaken an experience of our world and each other which is ethical, non-exploitative, compassionate and mutually enhancing.
— Lorin Hollander

To stop the flow of music would be like the stopping of time itself, incredible and inconceivable.

— Aaron Copland

Pulse Point

Maybe a gnat's nudge of the cheek recalls its short life, and ours. Maybe a table holds where it's placed. The documentary footage in which we live speeds fast forward, suspending weight in the indivisibility of life support and ice-capped tops of human thought.

Presence becomes its reason for being, its red-violet burst of indwelling meaning. Complexity along the fluid borders of flux will be layered plumb, when unity splits into parts of the moment approaching its pulse point in waking.

At first we were held in arms of the beautiful mother or father, and treated as people alive. We were carried as those who belong where inclination evolves, and complexity draws the world back into the brain with its places reserved for each being alive.

Tendering pours of torque have been intricate conditions of the species. Symbiosis lengthens in hammer-tongue gusts, in rib-root multiples grown out of gravity and standing.

Where we meet to negotiate our agreement, the table seems solid enough. The iron core of Earth holds us to the floor which was solid enough when we walked in.

Questions for Applicants (Form B)

How organic are you? How do you do? Are you quite green? Are you someone with a footprint in the darkness of man?

Do you compost yourself? How does loss of albedo strike you? Faced with a scarcity, do you naturally attempt to conceive? What causes you to experience more or less? Can't you swim? Do you work well in a fever? Will you drink from a well?

Where do you find yourself moving? Do you have a few moves? Do you own your own ounce of salt? Are you worth it? Do you keep tight bolts? Is your bean garden hoed?

How many fingers do you see? Haven't you been sad? Haven't you bet your life? Aren't you addicted to gambling? Where have your bees gone? Isn't the moment all we have? Will you say anything because that's how it is? Is that how it is? What frightens you most about disappearances of August moths?

Can you maintain a well-planned spontaneity as promotes trust? Once you've manufactured evidence, do you tend to stand by it? You don't pull your punches, do you? Are you a serious fucker? Do you have a will? Do you consider yourself well-concentrated?

Will you compose your own score and feather your nests? Where has your beer gone? Where have all the flowers gone? What have you done to effect organic change in the future? Do you clean up well? Do you compose yourself in the morning?

Reasoning from Before

Since we came to the planet for wild torches and nightcrawlers that reach mystically out of their tunnels to find each other, we were soon glad to see soft black ground the mushrooms have made of primal rock.

We arrived from various far-off systems of suns in starlit sky charted on periodic tables and learned to speak using words by hearing the light in voices. We've always seen ourselves in what has had its eyes looking.

As kids, squealing, taken over by some joy or aching, our bellies would be half filled, perhaps with apple, when we'd squat to revel in raw sound trans-mitted by breathing cells of ours that, growing us into shape, would let us grin or cry from inside out into mother air.

Romney and the Nature of Light

Light does and doesn't quite exist. A contiguous series of vacillations in the spectrum, it's an emergent property of infinitesimal disturbances in make-up of the time-space cosmos, an emanation of adjustments to matter which proceeds only so fast, at a frequency that depends on relative conditions. Light's emergence, of course, can be profound enough to inspire awe or terror in sync with a person's state and proximity.

Also in the mix would be residuals of childhood learning gone right or wrong, or shock to wherewithal which may have flooded the cells with chemical spills knocking out thought, and so on, with the brain stumped, asking *How could it have happened?* Or *Aren't we all relatives, with an ancient mother in common?*

Speaking of questions of genetic inheritance, how did candidate Romney end up taking credit for causing his effects, ignoring all he was given, the money and benefits of the community, not to mention long-term breakthroughs of evolution or labors and intelligence of anyone he's paid? When he was just a little Romney, what was he being fed?

There's this Romney and the human condition question, this Romney and seven-going-on-nine billion people question, along with inquiries waiting for takers on Romney and community necessity, Romney and Jeffersonian liberal education, Romney and mental constructs of identity, Romney and acting the opposite of Jesus in the New Testament, and so on.

The nature of light is to spread out, to give itself away wherever it goes. Whatever it is, a form of light exists within thinking, another kind within cells. The nature of light is to feed the world and then reveal it to itself, to keep going and keep thought going, where we're more the same than anything else we might say.

Saucer Fungus on a Doug Fir Stump by a Black Beetle

Harder than the wood it's ingested, the shelf fungus is a discus belonging to no one. On the half-standing stump, it can't stop glowing as if made to show through dark its rings within rings turning in slow motion to matter. So it sinks its roots to the mineral core and last rain drawn from Earth in the raven calm, through uncounted weeks of soil and moss designed by the cells.

The saucer fungus was a hunger at first, flying through night, through shades of a crow feather, the deep indigo in purple-black, night violet in white-black-black, where ravens and jays live in Pacific firs that escaped two-person saws, and grew into air from the underground. Spear-tip spores on thermals sailed from the event horizon to the beds of dying-down bark, to the rarest oils of stone from under many feet of needles and heavy sway of limbs.

Further spores jet in on the wind as thick-varnished beetles labor at the lip of salts and small harbors, in the mother language of serious digging, in descended light that lets them work, floating with heaviness. Day and night fail to stop. Beetles heat the light in their hulk. Lightning locks into breathing cells on the up-and-down wiring of air. Distance dissolves, vanishing as light shows from capillary ambers working.

Fast on the stump, shelf fungus echoes in rings within saucering over dust-quiet crawlers and worm-holing fibrous pods. Cells know what they see, where the scarlet-dark void has its billion trillions if it has one, on a path between rises to the lip of a tunnel.

Sayings as They Overflow

Worlds the sunflowers open will be overflowing on Earth, when it reaches velocity in time.

Flames have begun in an instantaneous part of the ambient Big Bang echoing *om*.

Taking a step carries along the past parents of parents, parents of remote great grandparents, back to more people alive than were counted for the medieval European census.

Where anything known approaches infinity, it's already been root-sinking plants feeding on light.

Back-country great horned owls will have bursts and brushes with blank dark extractions flying ahead and will be riding on shoulders of those who have found their backbones.

A North Atlantic right whale will inhale with the great celestial lungs of an ancient Celtic arena.

The exquisite Gorge will have waterfalls split by a drop of sudden rain recovering in a Stonehenge of rain.

Even a loose sprout driven by current appears to have old salmon shadows thrown onto many east county porches.

Scarcities

[*Information Found in Scientific American, August 2008*]

Around the world, the cost of food has doubled in three years. The Liberian cash needed for a bag of brown rice three years ago now buys a handful.

Switching what trucks and cars burn, from gasoline to hydrogen, would stop the flow of carbon pumped from the underground into air, but this is decades off.

Chemists have identified a substance that will block bitterness, fooling taste buds. Consider the possibilities, all that can be added at the factory to truly bitter foods.

Researchers have learned from lotus plants that small bumps on top of the leaves slope so water drops form and are instantly repelled.

Water drips from lotus leaves, carrying off any mud, wafts of curry, powders of incense. East Indians know that clean lotus plants may be growing in muddy ponds.

The Stenocara, an African desert beetle, has bumps on her back. She just lifts her back straight up, opens her mouth below, to drink mists from the air.

A new technological device for condensing moisture out of the air is being refined for those who don't have enough water.

Slowly Leaving Orbit

Hunger can make you unknowing and weak, unable to speak your father's language.

Forest floors still traverse lightning-first into the collective past. Racing on with evolutionary ancestry, the equator insinuates itself with a little of what we're forgetting.

Doesn't survival shape bones for the torso out of wild rain and the spectrum? The poles hold onto distant wheeling suns and planets. The moon's unfinished arc sharpens over time into the blade on a North African sickle before vanishing. Now when it's dark, it's blank, and prophesying ends.

The moon circles over this place it was followed many years for what it told us. It shows up mostly unnoticed, slowly severing connections with Earth.

Many around here deny they know nearly a billion people on Earth are suffering in utter poverty, lacking regular food.

Seismic heat radiates from the slightest cracks in molecular drive.

Heart-pulse blood's swimming new breath to every cell alive.

Hasn't fresh bread and its absence advanced unfinished history? Haven't lower and upper zero flooded mathematical propensity, where running cities on sunlight could happen as a breakthrough in integrity?

Small Place

Inexhaustible mystery may have been one of the more fundamental concepts helping ancestors survive. Past certain points, what else would do?

At the fairgrounds, rigged-up generators snoring and rattling current into searchlight caldrons still cast incandescent beams thicker than ancient redwoods into the post-WWII night. Across the way, engines of the '50s kick in, roaring with elemental calliope bursts, electrons leaping between valences, as fireflies throw master switches on and off head-high when you're a kid.

In the museum of acquisitions, emptiness burns in a golden bowl.

Bull's horns of *yes* or *no* drill into nakedness, with nothing to lose.

Prehistoric single-line paintings of mammoth remain avant garde.

But what's going on, when intentionally embracing ignorance, an elected official shrugs? Looming consequences, hard to pinpoint without research equipment, have shock, rage, crazy gambling, pea-green sickness, and flat-out scarlet harm racing off for the most vulnerable in proportion to the extent the planet's compromised or reasoning's sabotaged. No one's sure what'll happen.

Agitations crackle over wide-band radio tuned to emergency frequency in the kitchen where a friend's deputized father pours coffee from his thermos, waiting with battlefield readiness to rush to the next bad accident. His Civil Defense hardhat hangs on a hook. His all-purpose rig's equipped with ropes, chains, bandages, tourniquets, wrenches, blankets, and so on.

Serious accidents occur, where mass lifts within swells of bearing and awareness reaches for more.

Summer Day

At the first intersection past the river, a woman with peppery spiked hair herds a little boy through the heavy door of the piano building next to the absolute theater.

On the sidewalk by the wooden restaurant and white brick shop of found objects, the slender guy in a bright blue helmet lifts it off of the complexity which is his. He stands, looking across the land of scars on the avenue, helmet under his arm, his shoulders square and stance planted, as if the battle were long before and the harvest coming in strong.

In shimmering solar energy, young maples are scrubbing their share of the soup-pot sky when an anonymous European car passes, revealing the exquisite copper face in profile, an Andean woman, steering in a direction far away and yet present.

The vehicles of this moment in eternity have been stopped then started in a bumper-to-bumper crawl over the river of therapeutic bridges.

Halfway across the cement bridge, a man who obviously was kidnapped at an early age and repeatedly tortured in a barbershop chair has been walking east, beefy forearms at his chest, elbows slowly moving up and down at his sides as he's cracking his knuckles, walking in a field of energy, where it won't be long before he's broken out his raven's wings and flying home to Montana.

The Ballpark of Exchanged Gnosis

When the bat strikes light, small boys can be lifted, if only from a taste of the future not going away for a while.

The pitch of swaying crowds in pre-Columbian sun raises hair on the arms of outfielders hoofing the ground, making it part of themselves, whatever they end up doing.

The shortstop who's quick enough to cover more ground than anyone works his glove that gives off scent of the leather bag of the doctor who was ready for anything he came across on house calls.

The catcher, having sacrificed his sense of safety from early on, squats in the middle of danger that only makes him more sure of himself, curing and rubbing more life into the bull's eye mitt which long ago became part of him.

The past exists when a baby cries or when jars in apartment kitchens are unscrewed by hands the size of ball gloves, as when nitrogen rings numbly over the dirt where infielders lean slightly ahead into the pitch until the broad air stops holding it off from smacking into the catcher's mitt beyond question.

Each time the ball rolls off the pitcher's fingertip torque, the future wavers in a heat mirage, until what happens starts closing its clamshell over it. The batter swears to an invisible coach, *It was a spitball, slipping out of a curve while the center moved, and I couldn't see what was coming.*

When the moment arrives, as the ball's hit, the people racing around the diamond will never be only themselves. A few thousand people are rounding first, then sliding into second in a demonstration of political will.

A few thousand more are signaled around third while hundreds play right field, angling into the bounce, bare-handing the ball and firing it at the heart of the catcher, who refuses to apologize for his brawn by denying science.

The baserunner, no longer forced into keeping away from home, suddenly becomes twenty thousand leaping in a twist, half sliding into the unknown, risking everything, maneuvering like nobody else.

The Cat

The black-and-white cat dreams through a forest of night lit by wind of her body. The scent of ground and outside air pours over her.

Sleeping, she turns back through energy of her waking. She's like a person sleeping and dreaming, except she's curled her bones around breathing.

Whatever has gotten us this far, however much procreation may have come to term or whatever cellular history carries us as we walk, our birth animal keeps us alive, tracking the presence of overhead sun.

We accept the sleep entrusted to us as if we were going home. In the long run, we know our lives are part of biological grace that gave us shape over eons and woke within us what might have remained unknown.

The Ends of Desire

Sun branches through cells in almonds and shade, as through the shock of hips caught in a soundless ring of mud-caked Ice Age bells or muscular Brazilian canopies dragging out their disappearance. When quick heel clicks echo in marble basements downwind from anyone's vaults of lessers and betters, the stealing away in night-blinding muds commences.

Ennui polices socioeconomic succession with flat-out torque at the core of coal-swollen externalities. Agility sways, as deep blue intensifies behind blood-bearing flags that followed Magellan around the global horn.

Feathering off melts in the saucering galaxy, in an overflow is root heart heat in a roar. Where hard-wired displays of impermanence shatter around animals from before words, the blue bowl holds open for steaming soups, or the brush of tiniest mineral ribs of a cell.

The Future of Rhino Horn

The evolution of finch wings continues, where the moment resembles genetic complexity. A scent of topsoil spills through the air, where whispered voices in front of art may be turning through what they've received.

Mammoth with prophecy and primitive belief, the idea of terrestrial endlessness results in business as usual. In the future, those born into the day's work will find their hands full. As the coliseum of descendants fills its chest, the beauty of finch feathers will reveal what they've learned. Working the perimeter will be multiple aortic regenerations rippling along transpolar arcs the moment in which being someone means the same being for all species.

Where Mennonite lines of ancestors help deploy these eyes, slow-rising Celsius reels at the lip of vulnerability. As always, catalytic rhino horn will only work when left where it belongs.

The Idea of 2020

The laboratory door opens on Vajrayana emptiness. Any number of thoughts may be firing the same moment, some that may have been looking ahead to this day when they were just little risks to the future, long before the clear-cut Amazon under flat and deep overhead sun.

Exhalations of iridescent beetles and under-dirt badgers are mixed within breath, released as they were in trust of the place where we've found a heat-pulse chance.

Core-spun seedlings sweeten in elastic time, in the prisming spectrum, where sun rises in leaves, the unknowable future making the place smaller.

This must be where the compass needle points to the sun, where taste of the air and water speaks to the living cells. For greater needs than birth have been saucering out of the gene pool, as this era of global hot spots launches out through time.

On the honey-hive prairie, innocence and inception continue. But the place no longer belongs to old assumptions of endlessness, and the present revolves on its sunken taproot of jaguars.

The Idea of 2034

For thirty minutes, the climatologist explained his calculations of Arctic feedback mechanisms, citing top sources. Finally, I think, the media reveals urgency of the era and import of the IPCC and doesn't preemptively implode under atmospheric pressure of corporate control.

Relieved by the directness, we leaned into detail of clathrate releases of methane around sea ice ... *catastrophe ... methane observed now bubbling up around sea ice... could lead to a massive burp of it ... ruining conditions needed by plants ...*

He hammered a bronze Mayan gong the size of a person, in a molecular subdimension of my bodily cells. The hammer struck, the sun descended, and heaves of the ocean broke on banks. Metallic echoing probed the marrow of future buried bone. Shimmering with the unseeable, the gong shuddered and sunk under sea level until outside and inside stopped.

Waves of denial, panic, hostility, bursts of depression, bouts of explosive hedonism, maladaptive court-case splurges of casino and involuntary fasting, witchdoctor bargaining, sea-floor spudding, gnawing on rolls of loose teeth and gravel mouse droppings with bones of the skull, you name it, spread between disrupted symbiotic colonies of other species sharing present neuro-dynamic locomotion of these cells.

Immersed in life-support adaptation issues, I still wasn't ready to hear the process suddenly has been speeding up and could pass a point beyond which leafing green species in the northern hemisphere will be unable to survive—in maybe twenty years his numbers show, which is to say 2034.

The Idea of 2039

Hendrix jamming at the edge of volcanic expansion in the '60s skips ahead to 2039 night-swimming through the fossil-flamed continuum. As crow wings open and close over more than we know of loss or gain, the collaborative intensity of cells revolves around food enough for the billions of us alive.

Where the job's been taking you, fresh-baked bread already exists. A chance to live arcs between poles. A muffled hand-bell can be ringing in the ancestral belly of mercy, when being in the present multiplies within being.

How far back does the mind go? When earliest cells began to collaborate on the species, how did they decide? As spiral determination asserts impulse along meridians, the ancestral body's shape circulates with contemporary communications between cells.

Could instrumentation of a high order counter the chaos of runaway clashes of paradigm? Doesn't transnational profit billow over the run-off scoured fertilized seas? Won't international urbanity be flying in on a shoestring? Does the shock left over from birth help preserve humility, vulnerability?

If the eggshell Earth in 2039 isn't lit with a renaissance, the sun around here will be a burst or two harsher, with fewer atmospheric flocks. But the lightning in museums will still be coming from spiral engines of cells.

Can being in the present multiply further? When whole numbers stand face to face with the night sky, greater needs than birth will be dreaming up tough lessons for anyone trying to trust her ancestral inheritance of cells.

The Idea of Amplified Guitar

Driven by thunderhead convergence in the collective, the electric guitar hurls itself into the history of labor. It crashes into a hammered wall of extinctions and matriculates into peaks and life at sea.

Through photovoltaic bursts, it maneuvers around anthracite burns in the mouth of starlight, scientific vectors swiveling at the root of inception.

The guitar looks up from the medieval war, asking *What did I do?* It draws disorder out of suffering and fear. In the womb-swum communion of chances, it tenders electrical nerve, swaying with light through honey-hive prisms before rejoining the whole.

The last well-aimed guitar breaks up in suggestion. It comes back, drilling with nuclear searchlight the inability of air to utter a phrase without melting the poles.

The least touch of a live string can send out insistence, the longing or ache it has floating, resounding on waves that open through air with the guitar's refusal to cower before present conditions.

The plugged-in guitar is part of the modern human body which doubles as a ritual instrument of blessing and maybe surrender. Cypress-lit with ionospheric holds, it resonates within body, releasing it within heart-pulse circulation as aging Cold War silos stay closed.

The Idea of Bees

The project of hives is to keep the genome reaching into sunlight, attending the inside apparatus of beauty.

Where bees fly on their map, they end up burrowing in, lightly, electrically, touching what's central, what will be.

When they're ready, suddenly they return through the invisible air, leaving gold trails in the molecular wilds as they move through what we're breathing, the breath shared in circulation that keeps us alive.

The sun rises over oceans and crosses the sky all day in cells. Light pulses into plants, drawing them out, and more of them out of the Earth, delivering more mind into the mind, wherever we're going, however much of the future may be forgotten.

The way birth has left us in this era, long into the story of people, the route home has planted its root in muscle and veins under the sun.

Bees thrive in all directions around us, pollinating every neighborhood from their parallel amber-gold world.

The Idea of Eating an Orange

The problem with eating an orange comes at first, after gentle penetration, not too deep, with the thumbnails.

Maybe at first you thought the beautiful outer orange would taste delicious, but learned about bitterness and how the rind rips unevenly, as if something were seriously the matter.

In the face, maybe it is, of human domination of the planet and waste of what would be sacred—by protecting itself, the orange demonstrates its intelligence.

So the orange is both attractive and smart, and refuses to suffer fools.

In small hands, it remains impenetrable and whole, a recognized part of immensity, or another reason to burst out crying for arms of the mother.

In older hands, once the bitterness is done with, one's thumb and fingers are free to lightly trace over the inner sections until they part, to run along roundness that fits within the mouth.

This may reveal seeds of the tree and sweet tang, unlike any other, which fuels the desire to live.

The Idea of Heat 14° Higher Than Usual All July

An era of hot spots, drought, and floods launches from the encircling equator and poles.

All July in 2010 Moscow, the heat remains 14° higher than usual. Crops fail, leaves brown, and Russia turns to the world food market as the night stars burn.

Drought holds onto Syria, farms in the east good for little, and when people appeal for assistance from the monarch, the place ends up in war.

Where heat far beyond normal strikes, it leaves behind more for the vultures, for the hyenas and hornets. Take 2011 in Texas, where many deaths of people are being reclassified with climate change in mind.

Shifting currents in winds and oceans deliver unexpected lifts and falls, as flaming derivative algorithms scan the river-fog split second in which a few massive concentrations of wealth in the world grow.

While coal-burning the future to incapacity appears a specialty of ignorance these days, so does obstruction of changes we must make. In the history of mercy, what happens next?

The way Earth looks from space, a place interconnected, smaller than ancestors thought, what happens next could suddenly appear.

The Idea of Intrinsic Worth

Outlined in bolt Rouault, genomes unfold where the present shatters in light and dust, half-seen wheels spinning, mineral arcs verging, microbes spiraling, stallion mushroom floating spores, as breathing breaks in a slow wave through the body.

The downslope of fossil oil rocks at sea level, in hour-tone blues of the continuum. Cumulonimbus combustions fan out with far greater costs to the future than maybe we fathom.

In the meantime, glacial accumulation continues to feed inexplicable summer rivers of pulse, where gravity fills its chest with international philosophy fervent and sweet as the house finches.

As impulse may be, as those born into the day's work may find, as the dark draw of trunks differs from darkest night, the future arrives when a young girl in Minneapolis brushes her long hair before a mirror that would be empty without her.

The Idea of John Rogers

In 1550s London, John Rogers contemplated the newly translated Bible, sharing his views on the street, until a congregation formed around him.

Not every phrase coming from his mouth adhered to views of the state church.

Queen Mary I, deciding they'd had enough problems without the menace of a charismatic heretic inciting a sizeable ring of Londoners to go to hell, interceded. Her Majesty's guard relieved the self-anointed Reverend Rogers of his liberties and delivered him to the Queen's custody.

At the appointed hour, the Queen's executioner, serving at the Queen's pleasure, exacted her command. John Rogers was ushered in shackles through the commons to a place where torch was touched to kindling which encircled John Roger's stake in a burn to engulf him, sparing no one's comfort, in the 1554th Year of Our Lord, the 14th of the second month.

Following this, once they were able, the congregation sailed to the New World, settling in New England in pursuit of livelihood and thinking for oneself. Dedicated to education of the individual, eventually they produced *The New England Primer* to introduce writing and arithmetic to school children and anyone wanting to learn.

Part of the text covers the story of John Roger's martyrdom. With the words is a woodcut that shows the reverend in flames, at the stake, flanked by executioners, his distraught wife, and their nine children who were spared.

The New England Primer was used for a hundred years in large numbers of New World public schools.

The Idea of Native Frogs

There go the frogs again, ratcheting up DNA. I'm walking to the car after working late, mulling over my business, this week the argument for intrinsic worth of each species, when the frogs have no intention of giving in. The accelerators of their rigs are close to the floor, and they're pushing harder, lowering the pedal tone, entering the burn of trees.

I don't know what they think they're doing, stopping thought at this late hour. A number of cells come up with *self-organization,* but I can't keep out of their pulse, the tree frogs back in their electrical branches and hypnagogic muds, so sure of their purposes they're croaking, singing out through the atmosphere beneath moonlight a little more than half undone.

I think I'm off, done with horns of the dilemma, finished listening to whole other worlds packed into a room, when the frogs go on, tightening and loosening bolts that hold down the waters. They're busy, socket wrenches recoiling, as their throats resound, pouring out more glasses of mud and swamp gas, their eyes irradiant glassy coal.

They're all around, but nowhere, camouflaged by being themselves, which is good. I'm not ready to entertain, an army pack of reference manuals on my shoulders and wooden fish crate of yellow newspapers from days each week of my parents' European war making me a little wobbly, an instinctual duck in the crosshairs of their fire power and unification plan which exists outside my grasp.

Whatever they're selling, I'll take some—*Gnat wings? I'll buy a bag* to get them off my back where silent night triggers their urgency, and mine. It's not that they're asking, but telling me what to do: *Pity the poor mouse of the field. Pity the baby wren.*

They're out, raising their pulse, adjusting the ambient mist and stems before the abstract brines of stars, as this is a night to be living.

The Intersection of Traffic and Light

Cottonwood seed-snow drifts down, carried by unseen current from the unsold land behind us. Cars wait at the intersection where red light burns in scattering soft releases of seed parachuting from cottonwoods in the floating genome. Messages streak through patched-together space as if this kelson of creation were no different than probability, as if revolving magnetic arms of the planet's iron core were bubbling out of subdimensions and emptiness within daylight that maps what seems to be chance in which cells living in the world support one another.

Cottonwood trees of the unknown future drift through the intersection carried by aesthetic thermal spin into the slide of a van door shut, the roll of a shopping cart over a cracked walk, the swipe of a Studebaker fin out of the '50s as the slightest intent parachutes in slow motion, erasing speeds, sinking root into the future of intersections where red lingers and cars charge on, burning the matter of extractions.

The atmosphere around the human body mostly remains invisible, as if not a lot about breathing has changed, as if anyone's untoward chemistry were naturally reabsorbed by subdimensions, and the overflow of people were just a cottonwood snow revealing smallest currents and spires of local thermals in the intersection of matter and space, where the mastodon in the room is this air that lowers its massive head to aim terrific corkscrew tusks at the immediate causes of extinction.

The Pacific Trash Vortex

Swerving out of a blank spot of the back story far from shipping routes, soured and ingrown as seawaters go acidic, the vortex swells into a gaping wound where throw-outs collect trapped by current as cities spill more than anyone sees ending in the planetary swirl expanding north and east of Hawaii.

As Greenpeace folks gill-net some of it out, most pitched hollow polymer parts remain, most produce sacks in a briny stew through which jellyfish must parachute, through what's packing bellies of seabirds until they can't swallow, what's making elegant harbor seals clogged with lacerated wrappings and locking caps from tour-ship malls where waves float yellow wrenches through handbag rips sordid and barnacled.

Pitched bottles float, leaking quaternary ammonium biocides into parts of trawler-net comb-throughs catching many sea-swimmers stuffed with syndiotactic polypropylene to be eaten by others dying there with single net-snagged dolphins unable to breathe.

The vortex revolves in current swirling as if over a drain, but nothing goes down that doesn't ride on depths, cartwheeling with lost handles of blue toys and grocery aisle debris torn into death jellies stuck with roiling thick tons in the throat and belly of the ocean.

The Rooster Is Nowhere If Not Awake

There he is again, declaring the presence of civilization over unclear forces, crying out in the open, crowing up sound of a chickeny crowbar to have at the windows of light.

Wasn't it noon not long ago? Weren't we already awake? Has someone died, giving him the throne?

Maybe he'd fallen asleep in the shadow of cars, and after opening his eyes, believed he'd left the roost behind. And who wants to exist in loneliness at the edge of who knows where?

Again, he's throttling up, reporting his remote location and compliance, divining his own countervalence out of blood-red marrow, knowing what holds or goes off gray might need a push.

Is this the big chance we've been waiting for? Has he just cracked the code? Is this the first day of world peace? Is he calling out to the hens, and the forgotten mother of his fathers and mothers back through time?

The Same Rain

The rain flies in from Tibetan sanctuary. From Agincourt and Pacific rocks in the mist. Through onrushes of thick planning, it swells over the tractor rolling off on head-high Protestant wheels.

The future president notices the rain, each molecule with wind in it blowing in gusts over the sealed hotels. Raw economic fish axle through a last-century woodcut East European sky in a frame of fathery skeletons. Golden Rule girlhood sweeps over the roof in rain showering, shaping its algebraic grafts of arrival and escape.

The Declaration of Independence obfuscates the chatter in fields of attorneys, lost voter registrations in a dumpster, the medieval brick-street carnival in a chancellery, gravity strapped to a person's back not far removed from so many Cherokee coughs in space, from new loaves of bread suddenly sliced.

In the rain, this will be falling, much more than this, the coal-black philosophic flames thrown to the stone floor, what was lost looking at the human clock with colors of hair in sunlight, dust wheeled up behind giant John Deeres turning moist, the red-brown mane of late afternoon, phosphorescent bodies buoyant on rain.

The Scent of Afternoon Pavement

The anthropogenic era following high noon has people out walking to neighborhood shops, waiting for buses in the shade by fronts of '50s brick buildings, people parallel-parking dark vans on somebody's nickel, as Yakima apples fill on their root in the genome, and the burning moment of cars and trucks crawls along.

A small boy places his hand on the old brass handle of a door he can't open. After a tall woman with long half-gray hair whispers to him, he steps back, as if he's realized how small he is here, how large and unknown so much of the place remains.

When the shop door opens, bells ring next to a Nepalese ox collar in back of a silver-green manikin sporting a vest of woven neon tubes glowing like uranium in a breeder reactor. The objects for sale have come from remote locations where the combustion engine ended up going with New World money and state-of-the-art arms.

Air in this part of the city has a scent of charred meat with tinctures of Indonesian sandalwood and sanded wooden joists, but as if descendants have been opening Bulgarian travel trunks into pools of old-time shade.

Basements of recently torn-down houses are still steaming with a slow smoke of aging cardboard boxes, with nutmegs and extraordinary molds, as the spectrum of past chances prisms around them.

The Scent of Brown Rice

The present thunders in rings around ancient inheritance.

A mathematician on the sun-soaked train adjusts his breathing mask and thinks of his mother's hands.

Pre-symbolic machinery flourishes in mother leaves.

Ripened rain has nothing in mind but falling, as if to say we wouldn't want to be like it, drawn only toward the molten core of the planet, when desperately needed in so many places at once.

Forests thrive on sub-visible symbiosis of entities interconnected with multi-temporal fates. Absence and presence split and combine, extending through high-powered nakedness of surviving cells, as steaming brown rice blends with the shelter and offshore sway of the genome.

Through combustion in cells, ordinary matter forgives, as brown rice merges past splits at the edges of what we might or might not need.

The Unusual Medieval Occurrence

Music of the spheres has both genders, and speaks to both, lifting the mind. Many men mind Our word, and Our armies of horses will deliver a blessing as We are Queen.

Where the husband walks his hat down the lane, he is the many men brandishing upright shape. When the husband coughs, the many men who cough may sadden him or draw him on.

Where We are Queen, in Our presence are the many men and many women as may be. We will now commence Our project of unifying high and low.

All who may be here may know it here.

Should music be stirring here, with dancers or empty drift in time, a next plague, spell of heat, a known or unknown offspring—We will drum horses up from the south to the north, if We are King.

We will decide and judge within Our court, or cavort as We will, if We are King. And yet We are Her Majesty the Queen. We welcome all who hear Our declaration now of the end of hunger. Let fresh water be a province of all.

Any phrase having passed the lips Our mother and father bequeathed to us becomes law. We will now pity the little wife and husband, and offer them mercy.

Eradication of poverty around the Earth will proceed.

Tuesday at Work

Maples grow heavy with rain to come, thick with papery contracts missing from the 2062 files. Light within leaves crests in up-and-down waves, the huge vibrato leaving behind what couldn't simply merge.

Maybe you notice it more once you're alive, the integrative pulse within Zen holes in matter, as in puritans up-ending severe prayer in red wavering viscosity.

Tuesday rips into fresh card decks, shuffling families and mansions constructed around the yellow neon word NO. Photochemical contusions racing in compounds, Tuesday spreads, reverberating the size of continental drift, the first of morning bowling its open-source sphere into long afternoon.

Quick hours of day and night camphor the Cartesian grid.

Fast handling swings the glass doors open on the ends of vibration.

Trains on magnetic tracks hover beyond the lip of the event horizon.

Mel from Toledo would still be leaving the tire warehouse in locked-up 1968, his false teeth on the poker room floor, the whistle going out and coming on where the young are asking for work under the melting poles.

Roaming infrared around the clock, the virginal disco unrolls, the still point axle revolving, pieces of light sweeping through platitudes and plenty that continues or stops, as long as it takes.

The next quake through the cards speeds a few feet over the sea, brown pelicans fishing, when singly, stopped mid-air, one then a next strikes surgically bill-first in a sun-blaze where their jobs are clear.

Unscrolled Historical Volts

In 2007 the actuarial worth of a human being was set at $6.9 million by experts with much experience in that area. Valuable in insurance circles, the brain may be a nervous system that reconstructs dusk or dawn rather than taking them in head-on from a partially known, partially unknown world.

Divided by zero and the overpopulation, hundreds of thousands may be swimming in sea rain where every wheat plant and gnat sits on a vast foundation, as back-country apple-lit hills have gone buoyant in melts of reason guided on by the remarkable unseen bees.

If you've lost, say, the hoarse call of the great blue heron from along the Willamette after ten more years, what will you have in ten more? Certain coves have waves that will break anything they might have caught behind the recent sedations from constant airport alarm, engine blurs, tube blurts, and dried hair a few years after we've gone into flood-downs at the naked outskirts of unpaved '50s rummagings.

To be sure, the cure and the wound in the brain for which it was designed can hammer people into a corner, with penciled-in cheekbone petrifications taken to rustling unplugged then membranous as Bible pages ripped out. Volts across the expanse leave a few years floating in shafts.

So would individual death prove more expensive than joint death, or more than a whole group of people losing their lives together? Climate disruption could have us hungry and suffering, able to deny pounds of regret, and lit by questionings, crusted from summer and mule mud stomped in chambers of split-second burns of an eyeball, walking down drier needle-sponged trails with parts of the conscious mind still under construction.

Waiting for Thunder

Look at that lightning shuddering in the south, then nearly overhead: was there something you wanted to say? Each second is a stomach tightening, releasing, waiting on the cusp of scientific discovery, given this age.

And yet this age, too young to be left alone for long, too young to take its own vulnerability seriously, might be aging. Where a sensory unit slows to take in nectar from wisteria, its lightning turns transparent, into tiny flies, a native bee, a green hummingbird.

Inhaling after the flash felt in the chest, the human brain may be slow to exhale, to inhale, to undertake 360°, waiting for the tomato to ripen, for the beloved to make the call, for the tornado to rip the house away, for neocortical ignition to fire waking and sleep.

Maybe we're waiting for the concept of recycling to reach actual Ohio, or for the snow of pollen, numerical and ignited, to bring honeybees back in the gap between light and sound, wanting to hear the lightning flash finally complete its shuddering of the sky, to release us from not knowing, in the cloud cortex of future rain.

We Must Adapt

Solar immensities in leaves and microflora have, of course, kept human hunger alive. Seven billion are quickly becoming eight, then nine approaching ten around 2050, unless hunger sets us back.

Each of us has similar needs, and at least archaic mothers in common, though we forget.

Through the lift or fall of electric nerve, root threads under slow-motion bearing of one to the next, the scarlet *no return* of philosophical doors, collisions have been written across the face of the sun—

The wall of sleep suffers from burst-horse coal.

Admiral teeth decorate the uniform ocean floor.

Torches pave a wingspan path from birth.

Affinity, unfinished, undergoes the longer term.

Emily Carr's inexorable day-lit spiraling galaxies turn through principalities in sea-bellowed blameless wind.

Where so many bodies wake or sleep, fresh loaves bake within genetic code. The double helix resonates into a next generation as the heat slides species out of sync, and how much hunger can the planet carry?

What the Wind Keeps Saying

Let's make this quick, says the wind. *I don't believe in* a stained-glass donkey or wishful clover leading to dollars, when everything parks a gunned car on the back of a molecule, if that's what it does.

It could be, now more than ever, where current in the river lights the room with weight that floats, where what people have brought along can end up a wreck: the sousaphone on shore with health care vouchers and chicken wire, the statue of Ulysses S. Grant with a Pullman car and Cuban espionage film, with algae splashes in macaroni bank papers bleeding through, and so on.

In the shadow of head-high Farm-All wheels, the muskrats continue to root, glowing dark with rusting drive shafts and axles, in musket muds and liquid surfaces of stone, even as everything breathes when anything breathes.

For I believe in the only one person at a time, and in the next only one person, then in the next only one who is an ecosystem made possible by ecosystems and prayer-beaded sweat-salted deliverance of cells.

Doesn't the heart-filling brain spire in Barcelona? Didn't the Buddha meditate as a human who must eat or die, the way he was born? Doesn't the dragonfly quickly live for free? Wasn't Buddha one who saw the cost of even honest breath is suffering—where much is taken for granted—one whose respect for being never ended?

Where All Morning the Climate Increasingly Turns

> When you talk to people about going to Greenland and being there with this great ice sheet cliff going up to the ice cap, with Inuit elders, who are crying because, they told me, "When we were young, we came here. It was very hard to come here, but we came here. And even in the height of summer, there was no melting of the ice." And as we stood there, the water was roaring down this ice cliff, and huge, great sheets of ice were breaking off. And it was terrifying. It was absolutely terrifying. And learning that if the entire ice sheet melted and the ice cap melted, the oceans of the world would rise seven feet, and half of the world would be under water. — Jane Goodall

With every cell tied to what hopes in a molecule, this present era of remote consequences shows how every bite appears to be taking it out on the wilderness. Whether wisdom bites at your boot heels or the planet keeps going behind red curtains in the embassy of opposites, the baroque wheeling of air and water around Earth, as oceanic and precise as it may be, has been archaic and yet modern as morning skin.

With hypnotized angers going about bamboozling political will of the masses, morning sinks its silver spoon into the next roar of winds and squealing apparatus of eighteenth century assumptions no one's dismantled.

So who knows the exact month, season, and hour divine intent kicks in, cleaning the place up after us? In our freedom, who among us hasn't been owned by more than her own desire, more than inflations of rains that feather off the more serious melts in our saucering galaxy?

Morning light continues to shatter into spreads of color in parallel with the ancestral mind, as tongues are unconditional animals mostly before words, when isn't it time to admit that little exists other than what's here? Wouldn't you say a clear and present threat to the population has been the population? Hasn't all night through the day been where Northwest firs are risking their limbs?

But morning shouldn't be asking us much, even when all that exists plays the genetic keyboard before a fair share of bystanders and numerous public displays of impermanence.

Will of the Mind Falling Asleep

Take my mineral pulse through the restorations of coastal marsh of global saltwater in which the bleed of light carries lumber plumb following Japan's tsunami turning in waves on the ruined axle of shape.

Take my anthill across yards to where planetary rain forest lungs work in humming iridescence, where my carbon sinks to the sea floor in leaves, giving what matters to the trunk and roots, the way it was for ancestors.

Show me where Earth exists in a future as in the present, when the sky's so vast we're ants in ancient amphitheaters as the Greek chorus paces and the masks of each act made of clay and heat of the sun hover outside time.

Share my coal with the locomotive yields of heat from block-long stacks of engines hauling further ball joints for limbs and cradles for hips down tracks of spreading temperate vectors.

Let me release the cells of this skin ready to fly with gnat wings or lift into branches at muscular lengths, where Pacific rainforests still standing on their beds let mushrooms thread neural links to the tiniest.

Fill my roadside muskrat torso with Vajrayana emptiness that collects beautiful water with fauna visible only through lenses made for love of learning what is.

Throw my papers into vats of atmosphere which the mind serves at the pleasure of blue meridians of whales and urban silts in canyon swells that the climate turns on its stem.

Ring me up in the warehouse of bones through which cardinal imperative bursts with entanglement, the way animal refusals to stay in sight remember torches at the perimeter of archaic transaction.

Look into my eyes in the face of cattle and eyes the raccoon uses when marauding with fingertip perception in the continuum, when hunger launches from uncountable microscopic species living symbiotically in marshes of the mammal body.

You Can Sit for Hours

You can sit for hours in the summer by the red blossoms in back, near the blackberries that grew through, and still miss seeing her.

She may have spotted the red in the distance, and watched it go almost ultraviolet, drenched in sunlight in immensity, before she flew in from the nest she made by tying strips of grasses and debris into knots with her beak, fixing them with spider filament, then padding the bed with lichen.

And still it's sudden, when she shows up at a flower, drinking red. In a flash she's finished already, her body pivoting mid-air as if she weren't doing anything, making the sound you can't hear as much as sense within your chest, the hum of her invisible wings, a shudder nearly gone in the ruin of light, the whole of her turned emerald, into a piece of ceremonial jewelry escaping the Pharaoh.

In time, she flies in as she has before, and faster than trying to see her living at the speeds she does, already she's gone back into the vulnerable atmosphere, having left behind everything in colors the opposite of hers.

Photo by Bill Siverly

ABOUT THE AUTHOR

Since the early 1970s, James Grabill's poems and prose have appeared widely in literary journals primarily in the US, but also in the UK, Canada, Turkey, Algeria, Austria, and Australia. In the 1980s, he was a graduate fellow in Colorado State University's writing program, where he earned an MFA. He taught writing at CSU, for the Oregon Writers' Workshop, at Clackamas and Portland community colleges, and for eight years in the Clackamas Accelerated Degree Program, before joining full-time faculty at CCC where he has offered a range of writing courses (in exposition, argument, poetry, creative nonfiction, and technical writing), literature (including Shakespeare for two years), and an interdisciplinary inquiry sequence in sustainability, coordinated with the CCC Sustainability Lecture Series for which regional experts have discussed climate science, marine biodiversity, colony collapse disorder and the role of regional pollinators, renewable energy systems, the triple bottom line, and much more.

In addition to hundreds of appearances in literary and environmental periodicals, Grabill has published seven books of poems, two books of essays, and two poetry chapbooks. His *Poem Rising Out of the Earth and Standing Up in Someone* (Lynx House Press, 1994) was awarded the Oregon Book Award for Poetry in 1995. Additionally, three volumes have been OBA finalists: *Through the Green Fire* (Holy Cow! Press, 1995) in creative nonfiction, *Listening to the Leaves Form* (Lynx House, 1997) and *An Indigo Scent after the Rain* (Lynx House, 2003) in poetry. Now semi-retired, Grabill channels his energy into writing projects and public presentations. He continues to teach the "Pathways to Sustainability" sequence, which has helped inspire the two volumes of environmental prose poems, *Sea-Level Nerve,* published by Wordcraft of Oregon in 2014 and 2015. A long-time resident of Portland, Oregon, he is available for readings and workshops.

Books by James Grabill

Sea-Level Nerve, Book Two (prose poems) Wordcraft of Oregon, 2015
Sea-Level Nerve, Book One (prose poems) Wordcraft of Oregon, 2014
October Wind (poems) Sage Hill Press, 2006
Finding the Top of the Sky (creative nonfiction with poems)
　Lost Horse Press, 2005
An Indigo Scent after Rain (poems) Lynx House Press, 2003
Lame Duck Eternity (wild poems) 26 Books chapbook, 2000
Listening to the Leaves Form (poems and prose poems)
　Lynx House Press, 1997
Through the Green Fire (creative nonfiction with poems)
　Holy Cow! Press, 1995
Poem Rising Out of the Earth and Standing Up in Someone (poems)
　Lynx House Press, 1994 (Oregon Book Award, 1995)
In the Coiled Light (poems) NRG chapbook, 1985
To Other Beings (poems) Lynx House Press, 1981
Clouds Blowing Away (poems) Seizure and kayak Press, 1976
One River (a reverie of poems) Momentum Press, 1975

For other titles available from Wordcraft of Oregon, LLC
please visit our website at:
http://www.wordcraftoforegon.com

Also available through Ingram's, Amazon.com,
barnesandnoble.com, powells.com
and by special order through your local bookstore.